# Natural Selection

## CHARLES DARWIN

A Phoenix Paperback

This abridged edition published in 1996 by Phoenix
a division of Orion Books Ltd
Orion House, 5 Upper St Martin's Lane, London WC2H 9EA

This abridged edition contains the whole of Chapter 3
('Struggle for Existence') and the first part of Chapter 4
('Natural Selection') from *The Origin of Species*,
first published in 1859. The extracts reproduced here are from
the sixth and final edition published in 1872.

Cover illustration: Blue and yellow macaw, by Edward Lear
(1812–1888), © Private Collection/Bridgeman Art Library, London

ISBN 1 85799 605 4

Typeset by CentraCet Ltd, Cambridge
Printed in Great Britain by
Clays Ltd, St Ives plc

# Contents

## Struggle for Existence

Before entering on the subject of this chapter, I must make a few preliminary remarks, to show how the struggle for existence bears on Natural Selection. It has been seen in the last chapter that amongst organic beings in a state of nature there is some individual variability: indeed I am not aware that this has ever been disputed. It is immaterial for us whether a multitude of doubtful forms be called species or subspecies or varieties; what rank, for instance, the two or three hundred doubtful forms of British plants are entitled to hold, if the existence of any well-marked varieties be admitted. But the mere existence of individual variability and of some few well-marked varieties, though necessary as the foundation for the work, helps us but little in understanding how species arise in nature. How have all those exquisite adaptations of one part of the organisation to another part, and to the conditions of life, and of one organic being to another being, been perfected? We see these beautiful co-adaptations most plainly in the woodpecker and the mistletoe; and only a little less plainly in the humblest parasite which clings to the hairs of a quadruped or feathers of a bird; in the structure of the beetle which dives through the water: in the plumed seed which is wafted by the gentlest breeze; in short, we see beautiful adaptations everywhere and in every part of the organic world.

Again, it may be asked, how is it that varieties, which I have called incipient species, become ultimately converted into good and distinct species, which in most cases obviously differ from each other far more than do the varieties of the same species? How do those groups of species, which constitute what are called distinct genera, and which differ from each other more than do the species of the same genus, arise? All these results, as we shall more fully see in the next chapter, follow from the struggle for life. Owing to this struggle, variations, however slight and from whatever cause proceeding, if they be in any degree profitable to the individuals of a species, in their infinitely complex relations to other organic beings and to their physical conditions of life, will tend to the preservation of such individuals, and will generally be inherited by the offspring. The offspring, also, will thus have a better chance of surviving, for, of the many individuals of any species which are periodically born, but a small number can survive. I have called this principle, by which each slight variation, if useful, is preserved, by the term Natural Selection, in order to mark its relation to man's power of selection. But the expression often used by Mr Herbert Spencer of the Survival of the Fittest is more accurate, and is sometimes equally convenient. We have seen that man by selection can certainly produce great results, and can adapt organic beings to his own uses, through the accumulation of slight but useful variations, given to him by the hand of Nature. But Natural Selection, as we shall hereafter see, is a power incessantly ready for action, and is as immeasurably superior to man's feeble efforts, as the works of Nature are to those of Art.

We will now discuss in a little more detail the struggle for existence. In my future work this subject will be treated, as it well deserves, at greater length. The elder De Candolle and Lyell have largely and philosophically shown that all organic beings are exposed to severe competition. In regard to plants, no one has treated this subject with more spirit and ability than W. Herbert, Dean of Manchester, evidently the result of his great horticultural knowledge. Nothing is easier than to admit in words the truth of the universal struggle for life, or more difficult – at least I have found it so – than constantly to bear this conclusion in mind. Yet unless it be thoroughly engrained in the mind, the whole economy of nature, with every fact on distribution, rarity, abundance, extinction, and variation, will be dimly seen or quite misunderstood. We behold the face of nature bright with gladness, we often see superabundance of food; we do not see or we forget, that the birds which are idly singing round us mostly live on insects or seeds, and are thus constantly destroying life; or we forget how largely 3 these songsters, or their eggs, or their nestlings, are destroyed by birds and beasts of prey; we do not always bear in mind, that, though food may be now superabundant, it is not so at all seasons of each recurring year.

## The Term, Struggle for Existence, used in a large sense

I should premise that I use this term in a large and metaphorical sense including dependence of one being on another, and including (which is more important) not only the life of the individual, but success in leaving progeny. Two canine animals, in a time of dearth, may be truly said to struggle with each other which shall get food and live. But a plant on the edge of a desert is said to struggle for life against the drought, though more properly it should be said to be dependent on the moisture. A plant which annually produces a thousand seeds, of which only one of an average comes to maturity, may be more truly said to struggle with the plants of the same and other kinds which already clothe the ground. The mistletoe is dependent on the apple and a few other trees, but can only in a farfetched sense be said to struggle with these trees, for, if too many of these parasites grow on the same tree, it languishes and dies. But several seedling mistletoes, growing close together on the same branch, may more truly be said to struggle with each other. As the mistletoe is disseminated by birds, its existence depends on them; and it may metaphorically be said to struggle with other fruit-bearing plants, in tempting the birds to devour and thus disseminate its seeds. In these several senses, which pass into each other, I use for convenience' sake the general term of Struggle For Existence.

## Geometrical Ratio of Increase

A struggle for existence inevitably follows from the high rate at which all organic beings tend to increase. Every being, which during its natural lifetime produces several eggs or seeds, must suffer destruction during some period of its life, and during some season or occasional year, otherwise, on the principle of geometrical increase, its numbers would quickly become so inordinately great that no country could support the product. Hence, as more individuals are produced than can possibly survive, there must in every case be a struggle for existence, either one individual with another of the same species, or with the individuals of distinct species, or with the physical conditions of life. It is the doctrine of Malthus applied with manifold force to the whole animal and vegetable kingdoms; for in this case there can be no artificial increase of food, and no prudential restraint from marriage. Although some species may be now increasing, more or less rapidly, in numbers, all cannot do so, for the world would not hold them.

There is no exception to the rule that every organic

being naturally increases at so high a rate, that, if not destroyed, the earth would soon be covered by the progeny of a single pair. Even slow-breeding man has doubled in twenty-five years, and at this rate, in less than a thousand years, there would literally not be standing-room for his progeny. Linnaeus has calculated that if an annual plant produced only two seeds – and there is no plant so unproductive as this – and their seedlings next year produced two, and so on, then in twenty years there would be a million plants. The elephant is reckoned the slowest breeder of all known animals, and I have taken some pains to estimate its probable minimum rate of natural increase; it will be safest to assume that it begins breeding when thirty years old, and goes on breeding till ninety years old, bringing forth six young in the interval, and surviving till one hundred years old; if this be so, after a period of from 740 to 750 years there would be nearly nineteen million elephants alive, descended from the first pair.

But we have better evidence on this subject than mere theoretical calculations, namely, the numerous recorded cases of the astonishingly rapid increase of various animals in a state of nature, when circumstances have been favourable to them during two or three following seasons. Still more striking is the evidence from our domestic animals of many kinds which have run wild in several parts of the world; if the statements of the rate of increase 6 of slow-breeding cattle and horses in South America, and latterly in Australia, had not been well authenticated, they would have been incredible. So it is with plants; cases could be given of introduced plants which have become common throughout whole islands in a period of less than ten years. Several of the plants, such as the cardoon and a tall thistle, which are now the commonest over the wide plains of La Plata, clothing square leagues of surface almost to the exclusion of every other plant, have been introduced from Europe; and there are plants which now range in India, as I hear from Dr Falconer, from Cape Comorin to the Himalaya, which have been imported from America since its discovery. In such cases, and endless others could be given, no one supposes, that the fertility of the animals or plants has been suddenly and temporarily increased in any sensible degree. The obvious explanation is that the conditions of life have been highly favourable, and that there has consequently been less destruction of the old and young, and that nearly all the young have been enabled to breed. Their geometrical ratio of increase, the result of which never fails to be surprising, simply explains their extraordinarily rapid increase and wide diffusion in their new homes.

In a state of nature almost every full-grown plant annually produces seed, and amongst animals there are very few which do not annually pair. Hence we may confidently assert, that all plants and animals are tending to increase at a geometrical ratio, – that all would rapidly stock every station in which they could anyhow exist, – and that this geometrical tendency to increase must be checked by destruction at some period of life. Our familiarity with the larger domestic animals tends, I think, to mislead us: we see no great destruction falling on them, but we do not keep in mind that thousands are annually slaughtered for food, and that in a state of nature an equal number would have somehow to be disposed of.

The only difference between organisms which annually produce eggs or seeds by the thousand, and those which produce extremely few, is, that the slow-breeders would require a few more years to people, under favourable conditions, a whole district, let it be ever so large. The condor lays a couple of eggs and the ostrich a score, and yet in the same country the condor may be the more numerous of the two; the Fulmar petrel lays but one egg, yet it is believed to be the most numerous bird in the world. One fly deposits hundreds of eggs, and another, like the hippobosca, a single one; but this difference does not determine how many individuals of the two species can be supported in a district. A large number of eggs is of some importance to those species which depend on a fluctuating amount of food, for it allows them rapidly to increase in number. But the real importance of a large number of eggs or seeds is to make up for much destruction at some period of life; and this period in the great majority of cases is an early one. If an animal can in any way protect its own eggs or young, a small number may be produced, and yet the average stock be fully kept up; but if many eggs or young are destroyed, many must be produced, or the species will become extinct. It would suffice to keep up the full number of a tree, which lived on an average for a thousand years, if a single seed were produced once in a thousand years, supposing that this seed were never destroyed, and could be ensured to germinate in a fitting place. So that, in all cases, the average number of any animal or plant depends only indirectly on the number of its eggs or seeds.

In looking at Nature, it is most necessary to keep the foregoing considerations always in mind – never to forget that every single organic being may be said to be striving to the utmost to increase in numbers; that each lives by a struggle at some period of its life; that heavy destruction inevitably falls either on the young or old, during each generation or at recurrent intervals. Lighten any check, mitigate the destruction ever so little, and the member of the species will almost instantaneously increase to any amount.

## Nature of the Checks to Increase

The causes which check the natural tendency of each species to increase are most obscure. Look at the most vigorous species; by as much as it swarms in numbers, by so much will it tend to increase still further. We know not exactly what the checks are even in a single instance. Nor will this surprise any one who reflects how ignorant we are on this head, even in regard to mankind, although so incomparably better known than any other animal. This subject of the checks to increase has been ably treated by several authors, and I hope in a future work to discuss it at considerable length, more especially in regard to the feral animals of South America. Here I will make only a few remarks, just to recall to the reader's mind some of the chief points. Eggs or very young animals seem generally to suffer most, but this is not invariably the case. With plants there is a vast destruction of seeds, but, from some observations which I have made it appears that the seedlings suffer most from germinating in ground already thickly stocked with other plants. Seedlings, also, are destroyed in vast numbers by various enemies; for instance, on a piece of ground three feet long and two wide, dug and cleared, and where there could be no choking from other plants, I marked all the seedlings of our native weeds as they came up, and out of 357 no less than 295 were destroyed, chiefly by slugs and insects. If turf which has long been mown, and the case would be the same with turf closely browsed by quadrupeds, be let to grow, the more vigorous plants gradually kill the less vigorous, though fully grown plants; thus out of twenty species growing on a little plot of mown turf (three feet by four) nine species perished, from the other species being allowed to grow up freely.

The amount of food for each species of course gives the extreme limit to which each can increase; but very frequently it is not the obtaining food, but the serving as prey to other animals, which determines the average numbers of a species. Thus, there seems to be little doubt that the stock of partridges, grouse, and hares on any large estate depends chiefly on the destruction of vermin. If not one head of game were shot during the next twenty years in England, and, at the same time, if no vermin were destroyed, there would, in all probability, be less game than at present, although hundreds of thousands of game animals are now annually shot. On the other hand, in some cases, as with the elephant, none are destroyed by beasts of prey; for even the tiger in India most rarely dares to attack a young elephant protected by its dam.

Climate plays an important part in determining the average numbers of a species, and periodical seasons of extreme cold or drought seem to be the most effective of all checks. I estimated (chiefly from the greatly reduced numbers of nests in the spring) that the winter of 1854–5 destroyed four-fifths of the birds in my own grounds; and this is a tremendous destruction, when we remember that ten per cent. is an extraordinarily severe mortality from epidemics with man. The action of climate seems at first sight to be quite independent of the struggle for existence; but in so far as climate chiefly acts in reducing food, it brings on the most severe struggle between the individuals, whether of the same or of distinct species, which subsist on the same kind of food. Even when climate, for instance extreme cold, acts directly, it will be the least vigorous individuals, or those which have got least food through the advancing winter, which will suffer most. When we travel from south to north, or from a damp region to a dry, we invariably see some species gradually getting rarer and rarer, and finally disappearing; and the change of climate being conspicuous, we are tempted to attribute the whole effect to its direct action. But this is a false view; we forget that each species, even where it most abounds, is constantly suffering enormous destruction at some period of its life, from enemies or from competitors for the same place and food; and if these enemies or competitors be in the least degree favoured by any slight change of climate, they will increase in numbers; and as each area is already fully stocked with inhabitants, the other species must decrease. When we travel southward and see a species decreasing in numbers, we may feel sure that the cause lies quite as much in other species being favoured, as in this one being hurt. So it is when we travel northward, but in a somewhat lesser degree, for the number of species of all kinds, and therefore of competitors, decreases northwards; hence in going northwards, or in ascending a mountain, we far oftener meet with stunted forms, due to the directly injurious action of climate, than we do in proceeding southwards or in descending a mountain. When we reach the Arctic regions, or snow-capped summits, or absolute deserts, the struggle for life is almost exclusively with the elements.

That climate acts in main part indirectly by favouring other species, we clearly see in the prodigious number of plants which in our gardens can perfectly well endure our climate, but which never become naturalised, for they cannot compete with our native plants nor resist destruction by our native animals.

When a species, owing to highly favourable circumstances, increases inordinately in numbers in a small tract, epidemics – at least, this seems generally to occur with our game animals – often ensue; and here we have a limiting check independent of the struggle for life. But even some of these so-called epidemics appear to be due to parasitic worms, which have from some cause, possibly in part through facility of diffusion amongst the crowded animals, been disproportionally favoured: and here comes in a sort of struggle between the parasite and its prey.

On the other hand, in many cases, a large stock of individuals of the same species, relatively to the numbers of its enemies, is absolutely necessary for its preservation. Thus we can easily raise plenty of corn and rape-seed, &c., in our fields, because the seeds are in great excess compared with the number of birds which feed on them; nor can the birds, though having a superabundance of food at this one season, increase in number proportionally to the supply of seed, as their numbers are checked during winter; but any one who has tried, knows how troublesome it is to get seed from a few wheat or other such plants in a garden: I have in this case lost every single seed. This view of the necessity of a large stock of the same species for its preservation, explains, I believe, some singular facts in nature such as that of very rare plants being sometimes extremely abundant, in the few spots where they do exist; and that of some social plants being social, that is abounding in individuals, even on the extreme verge of their range. For in such cases, we may believe, that a plant could exist only where the conditions of its life were so favourable that many could exist together, and thus save the species from utter destruction. I should add that the good effects of intercrossing, and the ill effects of close interbreeding, no doubt come into play in many of these cases; but I will not here enlarge on this subject.

## Complex Relations of all Animals and Plants to each other in the Struggle for Existence

Many cases are on record showing how complex and unexpected are the checks and relations between organic beings, which have to struggle together in the same country. I will give only a single instance, which, though a simple one, interested me. In Staffordshire, on the estate of a relation, where I had ample means of investigation, there was a large and extremely barren heath, which had never been touched by the hand of man; but several hundred acres of exactly the same nature had been enclosed twenty-five years previously and planted with Scotch fir. The change in the native vegetation of the planted part of the heath was most remarkable, more than is generally seen in passing from one quite different soil to another: not only the proportional numbers of the heath-plants were wholly changed, but twelve species of plants (not counting grasses and carices) flourished in the plantations, which could not be found on the heath. The effect on the insects must have been still greater, for six insectivorous birds were very common in the plantations, which were not to be seen on the heath; and the heath was frequented by two or three distinct insectivorous birds. Here we see how potent has been the effect of the introduction of a single tree, nothing whatever else having been done, with the exception of the land having been enclosed, so that cattle could not enter. But how important an element enclosure is, I plainly saw near Farnham, in Surrey. Here there are extensive heaths, with a few clumps of old Scotch firs on the distant hill-tops: within the last ten years large spaces have been enclosed, and self-sown firs are now springing up in multitudes, so close together that all cannot live. When I ascertained that these young trees had not been sown or planted, I was so much surprised at their numbers that I went to several points of view, whence I could examine hundreds of acres of the unenclosed heath, and literally I could not see a single Scotch fir, except the old planted clumps. But on looking closely between the stems of the heath, I found a multitude of seedlings and little trees which had been perpetually browsed down by the cattle. In one square yard, at a point some hundred yards distant from one of the old clumps, I counted thirty-two little trees; and one of them, with twenty-six rings of growth, had, during many years tried to raise its head above the stems of the heath, and had failed. No wonder that, as soon as the land was enclosed, it became thickly clothed with vigorously growing young firs. Yet the heath was so extremely barren and so extensive that no one would ever have imagined that cattle would have so closely and effectually searched it for food.

Here we see that cattle absolutely determine the existence of the Scotch fir; but in several parts of the world insects determine the existence of cattle. Perhaps Paraguay offers the most curious instance of this; for here neither cattle nor horses nor dogs have ever run wild, though they swarm southward and northward in a feral state; and Azara and Rengger have shown that this is caused by the greater number in Paraguay of a certain fly, which lays its eggs in the navels of these animals when first born. The increase of these flies, numerous as they are, must be habitually checked by some means, probably by other parasitic insects. Hence, if certain insectivorous birds were to decrease in Paraguay, the parasitic insects would probably increase; and this would lessen the number of the navel-frequenting flies – then cattle and horses would become feral, and this would certainly greatly alter (as indeed I have observed in parts of South America) the vegetation: this again would largely affect the insects; and this, as we have just seen in Staffordshire, the insectivorous birds, and so onwards in ever-increasing circles of complexity. Not that under nature the relations will ever be as simple as this. Battle within battle must be continually recurring with varying success; and yet in the long-run the forces are so nicely balanced, that the faces of nature remains for long periods of time uniform, though assuredly the merest trifle would give the victory to one organic being over another. Nevertheless, so profound is our ignorance, and so high our presumption, that we marvel when we hear of the extinction of an organic being; and as we do not see the cause, we invoke cataclysms to desolate the world, or invent laws on the duration of the forms of life!

I am tempted to give one more instance showing how plants and animals, remote in the scale of nature, are bound together by a web of complex relations. I shall hereafter have occasion to show that the exotic Lobelia fulgens is never visited in my garden by insects, and consequently, from its peculiar structure, never sets a seed. Nearly all our orchidaceous plants absolutely require the visits of insects to remove their pollen-masses and thus to fertilise them. I find from experiments that humble-bees are almost indispensable to the fertilisation of the heartsease (Violo tricolor), for other bees do not visit this flower. I have also found that the visits of bees are necessary for the fertilisation of some kinds of clover; for instance, 20 heads of Dutch clover (Trifolium repens) yielded 2,290 seeds, but 20 other heads protected from bees produced not one. Again, 100 heads of red clover (T. pratense) produced 2,700 seeds, but the same number of protected heads produced not a single seed. Humble-bees alone visit red clover, as other bees cannot reach the nectar. It has been suggested that moths may fertilise the clovers; but I doubt whether they could do so in the case of the red clover, from their weight not being sufficient to depress the wing petals. Hence we may infer as highly probable that, if the whole genus of humble-bees became extinct or very rare in England, the heartsease and red clover would become very rare, or wholly disappear. The number of humble-bees in any district depends in a great measure upon the number of fieldmice, which destroy their combs and nests; and Col. Newman, who has long attended to the habits of humble-bees, believes that 'more than two-thirds of them are thus destroyed all over England.' Now the number of mice is largely dependent, as every one knows, on the number of cats; and Col. Newman says, 'Near villages and small towns I have found the nests of humble-bees more numerous than elsewhere, which I attribute to the number of cats that destroy the mice.' Hence it is quite credible that the presence of a feline animal in large numbers in a district might determine, through the intervention first of mice and then of bees, the frequency of certain flowers in that district!

In the case of every species, many different checks, acting at different periods of life, and during different seasons or years, probably come into play; some one check or some few being generally the most potent; but all will concur in determining the average number or even the existence of the species. In some cases it can be shown that widely-different checks act on the same species in different districts. When we look at the plants and bushes clothing an entangled bank, we are tempted to attribute their proportional numbers and kinds to what we call chance. But how false a view is this! Every one has heard that when an American forest is cut down, a very different vegetation springs up; but it has been observed that ancient Indian ruins in the Southern United States, which must formerly have been cleared of trees, now display the same beautiful diversity and proportion of kinds as in the surrounding virgin forest. What a struggle must have gone on during long centuries between the several kinds of trees, each annually scattering its seeds by the thousand; what war between insect and insect – between insects, snails, and other animals with birds and beasts of prey – all striving to increase, all feeding on each other, or on the trees, their seeds and seedlings, or on the other plants which first clothed the ground and thus checked the growth of the trees! Throw up a handful of feathers, and all fall to the ground according to definite laws; but how simple is the problem where each shall fall compared to that of the action and reaction of the innumerable plants and animals which have determined, in the course of centuries, the proportional numbers and kinds of trees now growing on the old Indian ruins!

The dependency of one organic being on another, as of a parasite on its prey, lies generally between beings remote in the scale of nature. This is likewise sometimes the case with those which may be strictly said to struggle with each other for existence, as in the case of locusts and grass-feeding quadrupeds. But the struggle will almost invariably be most severe between the individuals of the same species, for they frequent the same districts, require the same food, and are exposed to the same dangers. In the case of varieties of the same species, the struggle will generally be almost equally severe, and we sometimes see the contest soon decided: for instance, if several varieties of wheat be sown together, and the mixed seed be resown, some of the varieties which best suit the soil or climate, or are naturally the most fertile, will beat the others and so yield more seed, and will consequently in a few years supplant the other varieties. To keep up a mixed stock of even such extremely close varieties as the variously-coloured sweet peas, they must be each year harvested separately, and the seed then mixed in due proportion, otherwise the weaker kinds will steadily decrease in number and disappear. So again with the varieties of sheep; it has been asserted that certain mountain-varieties will starve out other mountain-varieties, so that they cannot be kept together. The same result has followed from keeping together different varieties of the medicinal leech. It may even be doubted whether the varieties of any of our domestic plants or animals have so exactly the same strength, habits, and constitution, that the original proportions of a mixed stock (crossing being prevented) could be kept up for half-a-dozen generations, if they were allowed to struggle together, in the same manner as beings in a state of nature, and if the seed or young were not annually preserved in due proportion.

## Struggle for Life most severe between Individuals and Varieties of the same Species

As the species of the same genus usually have, though by no means invariably, much similarity in habits and constitution, and always in structure, the struggle will generally be more severe between them, if they come into competition with each other, than between the species of distinct genera. We see this in the recent extension over parts of the United States of one species of swallow having caused the decrease of another species. The recent increase of the missel-thrush in parts of Scotland has caused the decrease of the song-thrush. How frequently we hear of one species of rat taking the place of another 21 species under the most different climates! In Russia the small Asiatic cockroach has everywhere driven before it its great congener. In Australia the imported hive-bee is rapidly exterminating the small stingless native bee.

One species of charlock has been known to supplant another species; and so in other cases. We can dimly see why the competition should be most severe between allied forms, which fill nearly the same place in the economy of nature; but probably in no one case could we precisely say why one species has been victorious over another in the great battle of life.

A corollary of the highest importance may be deduced from the foregoing remarks, namely, that the structure of every organic being is related, in the most essential yet often hidden manner, to that of all the other organic beings, with which it comes into competition for food or residence, or from which it has to escape, or on which it preys. This is obvious in the structure of the teeth and talons of the tiger; and in that of the legs and claws of the parasite which clings to the hair on the tiger's body. But in the beautifully plumed seed of the dandelion, and in the flattened and fringed legs of the water-beetle, the relation seems at first confined to the elements of air and water. Yet the advantage of plumed seeds no doubt stands in the closest relation to the land being already thickly clothed with other plants; so that the seeds may be widely distributed and fall on unoccupied ground. In 22 the water-beetle, the structure of its legs, so well adapted for diving, allows it to compete with other aquatic insects, to hunt for its own prey, and to escape serving as prey to other animals.

The store of nutriment laid up within the seeds of many plants seems at first sight to have no sort of relation to other plants. But from the strong growth of young plants produced from such seeds, as peas and beans, when sown in the midst of long grass, it may be suspected that the chief use of the nutriment in the seed is to favour the growth of the seedlings, whilst struggling with other plants growing vigorously all around.

Look at a plant in the midst of its range, why does it not double or quadruple its numbers? We know that it can perfectly well withstand a little more heat or cold, dampness or dryness, for elsewhere it ranges into slightly hotter or colder, damper or drier districts. In this case we can clearly see that if we wish in imagination to give the plant the power of increasing in number, we should have to give it some advantage over its competitors, or over the animals which prey on it. On the confines of its geographical range, a change of constitution with respect to climate would clearly bestow advantage to our plant; but we have reason to believe that only a few plants or animals range so far, that they are destroyed exclusively by the rigour of the climate. Not until we reach the extreme confines of life, in the Arctic regions or on the borders of an utter desert, will competition cease. The land may be extremely cold or dry, yet there will be competition between some few species, or between the individuals of the same species, for the warmest or dampest spots.

Hence we can see that when a plant or animal is placed in a new country amongst new competitors, the conditions of its life will generally be changed in an essential manner, although the climate may be exactly the same as in its former home. If its average numbers are to increase in its new home, we should have to modify it in a different way to what we should have had to do in its native country; for we should have to give it some advantage over a different set of competitors or enemies.

It is good thus to try in imagination to give to any one species an advantage over another. Probably in no single instance should we know what to do. This ought to convince us of our ignorance on the mutual relations of all organic beings; a conviction as necessary, as it is difficult to acquire. All that we can do, is to keep steadily in mind that each organic being is striving to increase in a geometrical ratio; that each at some period of its life, during some season of the year, during each generation or at intervals, has to struggle for life and to suffer great destruction. When we reflect on this struggle, we may console ourselves with the full belief, that the war of nature is not incessant, that no fear is felt, that death is generally prompt, and that the vigorous, the healthy, and the happy survive and multiply.

## Natural Selection, or the Survival
## of the Fittest

How will the struggle for existence, briefly discussed in the last chapter, act in regard to variation? Can the principle of selection, which we have seen is so potent in the hands of man, apply under nature? I think we shall see that it can act most efficiently. Let the endless number of slight variations and individual differences occurring in our domestic productions, and, in a lesser degree, in those under nature, be borne in mind; as well as the strength of the hereditary tendency. Under domestication, it may be truly said that the whole organisation becomes in some degree plastic. But the variability, which we almost universally meet with in our domestic productions, is not directly produced, as Hooker and Asa Gray have well remarked, by man; he can neither originate varieties, nor prevent their occurrence; he can only preserve and accumulate such as do occur. Unintentionally he exposes organic beings to new and changing conditions of life, and variability ensues; but similar changes of conditions might and do occur under nature. Let it also be borne in mind how infinitely complex and close-fitting are the mutual relations of all organic beings to each other and to their physical conditions of life; and consequently what infinitely varied diversities of structure might be of use to each being under changing conditions of life. Can it, then, be thought improbable, seeing that variations useful to man have undoubtedly occurred, that other variations useful in some way to each being in the great and complex battle of life, should occur in the course of many successive generations? If such do occur, can we doubt (remembering that many more individuals are born than can possibly survive) that individuals having any advantage, however slight, over others, would have the best chance of surviving and of procreating their kind? On the other hand, we may feel sure that any variation in the least degree injurious would be rigidly destroyed. This preservation of favourable individual differences and variations, and the destruction of those which are injurious, I have called Natural Selection, or the Survival of the Fittest. Variations neither useful nor injurious would not be affected by natural selection, and would be left either a fluctuating element, as perhaps we see in certain polymorphic species, or would ultimately become fixed, owing to the nature of the organism and the nature of the conditions.

Several writers have misapprehended or objected to the term Natural Selection. Some have even imagined that natural selection induces variability, whereas it implies only the preservation of such variations as arise and are beneficial to the being under its conditions of life. No one objects to agriculturists speaking of the potent effects of man's selection; and in this case the individual differences given by nature, which man for some object selects, must of necessity first occur. Others have objected that the term selection implies conscious choice in the animals which become modified; and it has even been urged that, as plants have no volition, natural selection is not applicable to them! In the literal sense of the word, no doubt, natural selection is a false term; but who ever objected to chemists speaking of the elective affinities of the various elements? – and yet an acid cannot strictly be said to elect the base with which it in preference combines. It has been said that I speak of natural selection as an active power or Deity; but who objects to an author speaking of the attraction of gravity as ruling the movements of the planets? Every one knows what is meant and is implied by such metaphorical expressions; and they are almost necessary for brevity. So again it is difficult to avoid personifying the word Nature; but I mean by Nature, only the aggregate action and product of many natural laws, and by laws the sequence of events as ascertained by us. With a little familiarity such superficial objections will be forgotten.

We shall best understand the probable course of natural selection by taking the case of a country undergoing some slight physical change, for instance, of climate. The proportional members of its inhabitants will almost immediately undergo a change, and some species will probably become extinct. We may conclude, from what we have seen of the intimate and complex manner in which the inhabitants of each country are bound together, that any change in the numerical proportions of the inhabitants, independently of the change of climate itself, would seriously affect the others. If the country were open on its borders, new forms would certainly immigrate, and this would likewise seriously disturb the relations of some of the former inhabitants. Let it be remembered how powerful the influence of a single introduced tree or mammal has been shown to be. But in the case of an island, or of a country partly surrounded by barriers, into which new and better adapted forms could not freely enter, we should then have places in the economy of nature which would assuredly be better filled up, if some of the original inhabitants were in some manner modified; for, had the area been open to immigration, these same places would have been seized on by intruders. In such cases, slight modifications, which in any way favoured the individuals of any species, by better adapting them to their altered conditions, would tend to be preserved; and natural selection would have free scope for the work of improvement.

We have good reason to believe, as shown in the first chapter, that changes in the conditions of life give a tendency to increased variability; and in the foregoing cases the conditions have changed, and this would manifestly be favourable to natural selection, by affording a better chance of the occurrence of profitable variations. Unless such occur, natural selection can do nothing. Under the term of 'variations,' it must never be forgotten that mere individual differences are included. As man can produce a great result with his domestic animals and plants by adding up in any given direction individual differences, so could natural selection, but far more easily from having incomparably longer time for action. Nor do I believe that any great physical change, as of climate, or any unusual degree of isolation to check immigration, is necessary in order that new and unoccupied places should be left, for natural selection to fill up by improving some of the varying inhabitants. For as all the inhabitants of each country are struggling together with nicely balanced forces, extremely slight modifications in the structure or habits of one species would often give it an advantage over others; and still further modifications of the same kind would often still further increase the advantage, as long as the species continued under the same conditions of life and profited by similar means of subsistence and defence. No country can be named in which all the native inhabitants are now so perfectly adapted to each other and to the physical conditions under which they live, that none of them could be still better adapted or improved; for in all countries, the natives have been so far conquered by naturalised productions, that they have allowed some foreigners to take firm possession of the land. And as foreigners have thus in every country beaten some of the natives, we may safely conclude that the natives might have been modified with advantage, so as to have better resisted the intruders.

As man can produce, and certainly has produced, a great result by his methodical and unconscious means of selection, what may not natural selection effect? Man can act only on external and visible characters: Nature, if I may be allowed to personify the natural preservation or survival of the fittest, cares nothing for appearances, except in so far as they are useful to any being. She can act on every internal organ, on every shade of constitutional difference, on the whole machinery of life. Man selects only for his own good: Nature only for that of the being which she tends. Every selected character is fully exercised by her, as is implied by the fact of their selection. Man keeps the natives of many climates in the same country; he seldom exercises each selected character in some peculiar and fitting manner; he feeds a long and a short beaked pigeon on the same food; he does not exercise a long-backed or long-legged quadruped in any peculiar manner; he exposes sheep with long and short wool to the same climate. He does not allow the most vigorous males to struggle for the females. He does not rigidly destroy all inferior animals, but protects during each varying season, as far as lies in his power, all his productions. He often begins his selection by some half-monstrous form; or at least by some modification prominent enough to catch the eye or to be plainly useful to him. Under nature, the slightest differences of structure or constitution may well turn the nicely-balanced scale in the struggle for life, and so be preserved. How fleeting are the wishes and efforts of man! how short his time! and consequently how poor will be his results, compared with those accumulated by Nature during whole geological periods! Can we wonder, then, that Nature's productions should be far 'truer' in character than man's productions; that they should be infinitely better adapted to the most complex conditions of life, and should plainly bear the stamp of far higher workmanship?

It may metaphorically be said that natural selection is daily and hourly scrutinising, throughout the world, the slightest variations; rejecting those that are bad, preserving and adding up all that are good; silently and insensibly working, *whenever and wherever opportunity offers*, at the improvement of each organic being in relation to its organic and inorganic conditions of life. We see nothing of these slow changes in progress, until the hand of time has marked the lapse of ages, and then so imperfect is our view into long-past geological ages, that we see only that the forms of life are now different from what they formerly were.

In order that any great amount of modification should be effected in a species, a variety when once formed must again, perhaps after a long interval of time, vary or present individual differences of the same favourable nature as before; and these must be again preserved, and 31 so onwards step by step. Seeing that individual differences of the same kind perpetually recur, this can hardly be considered as an unwarrantable assumption. But whether it is true, we can judge only by seeing how far the hypothesis accords with and explains the general phenomena of nature. On the other hand, the ordinary belief that the amount of possible variation is a strictly limited quantity is likewise a simple assumption.

Although natural selection can act only through and for the good of each being, yet characters and structures, which we are apt to consider as of very trifling importance, may thus be acted on. When we see leafeating insects green, and bark-feeders mottled-grey; the alpine ptarmigan white in winter, the red-grouse the colour of heather, we must believe that these tints are of service to these birds and insects in preserving them from danger. Grouse, if not destroyed at some period of their lives, would increase in countless numbers; they are known to suffer largely from birds of prey; and hawks are guided by eyesight to their prey – so much so, that on parts of the Continent persons are warned not to keep white pigeons, as being the most liable to destruction. Hence natural selection might be effective in giving the proper colour to each kind of grouse, and in keeping that colour, when once acquired, true and constant. Nor ought we to think that the occasional destruction of an animal of any particular colour would produce little effect: we should

remember how essential it is in a flock of white sheep to

destroy a lamb with the faintest trace of black. We have seen how the colour of the hogs, which feed on the 'paint-root' in Virginia, determines whether they shall live or die. In plants, the down on the fruit and the colour of the flesh are considered by botanists as characters of the most trifling importance: yet we hear from an excellent horti-culturist, Downing, that in the United States smooth-skinned fruits suffer far more from a beetle, a Curculio, than those with down; that purple plums suffer far more from a certain disease than yellow plums; whereas another disease attacks yellow-fleshed peaches far more than those with other coloured flesh. If, with all the aids of art, these slight differences make a great difference in cultivating the several varieties, assuredly, in a state of nature, where the trees would have to struggle with other trees and with a host of enemies, such differences would effectually settle which variety, whether a smooth or downy, a yellow or purple fleshed fruit, should succeed.

In looking at many small points of difference between species, which, as far as our ignorance permits us to judge, seem quite unimportant, we must not forget that climate, food, &c., have no doubt produced some direct effect. It is also necessary to bear in mind that, owing to the law of correlation, when one part varies, and the variations are accumulated through natural selection, other modifications, often of the most unexpected nature, will ensue.

As we see that those variations which, under domestication, appear at any particular period of life, tend to reappear in the offspring at the same period; – for instance, in the shape, size, and flavour of the seeds of the many varieties of our culinary and agricultural plants; in the caterpillar and cocoon stages of the varieties of the silkworm; in the eggs of poultry, and in the colour of the down of their chickens; in the horns of our sheep and cattle when nearly adult; – so in a state of nature natural selection will be enabled to act on and modify organic beings at any age, by the accumulation of variations profitable at that age, and by their inheritance at a corresponding age. If it profit a plant to have its seeds more and more widely disseminated by the wind, I can see no greater difficulty in this being effected through natural selection, than in the cotton-planter increasing and improving by selection the down in the pods on his cotton-trees. Natural selection may modify and adapt the larva of an insect to a score of contingencies, wholly different from those which concern the mature insect; and these modifications may effect, through correlation, the structure of the adult. So, conversely, modifications in the adult may affect the structure of the larva; but in all cases natural selection will ensure that they shall not be injurious: for if they were so, the species would become extinct.

Natural selection will modify the structure of the young in relation to the parent, and of the parent in relation to the young. In social animals it will adapt the structure of

34

each individual for the benefit of the whole community; if the community profits by the selected change. What natural selection cannot do, is to modify the structure of one species, without giving it any advantage, for the good of another species; and though statements to this effect may be found in works of natural history, I cannot find one case which will bear investigation. A structure used only once in an animal's life, if of high importance to it, might be modified to any extent by natural selection; for instance, the great jaws possessed by certain insects, used exclusively for opening the cocoon – or the hard tip to the beak of unhatched birds, used for breaking the egg. It has been asserted, that of the best short-beaked tumbler-pigeons a greater number perish in the egg than are able to get out of it; so that fanciers assist in the act of hatching. Now if nature had to make the beak of a full-grown pigeon very short for the bird's own advantage, the process of modification would be very slow, and there would be simultaneously the most rigorous selection of all the young birds within the egg, which had the most powerful and hardest beaks, for all with weak beaks would inevitably perish; or, more delicate and more easily broken shells might be selected, the thickness of the shell being known to vary like every other structure.

It may be well here to remark that with all beings there must be much fortuitous destruction, which can have little or no influence on the course of natural selection. For instance a vast number of eggs or seeds are annually

devoured, and these could be modified through natural selection only if they varied in some manner which protected them from their enemies. Yet many of these eggs or seeds would perhaps, if not destroyed, have yielded individuals better adapted to their conditions of life than any of those which happened to survive. So again a vast number of mature animals and plants, whether or not they be the best adapted to their conditions, must be annually destroyed by accidental causes, which would not be in the least degree mitigated by certain changes of structure or constitution which would in other ways be beneficial to the species. But let the destruction of the adults be ever so heavy, if the number which can exist in any district be not wholly kept down by such causes, – or again let the destruction of eggs or seeds be so great that only a hundredth or a thousandth part are developed, – yet of those which do survive, the best adapted individuals, supposing that there is any variability in a favourable direction, will tend to propagate their kind in larger numbers than the less well adapted. If the numbers be wholly kept down by the causes just indicated, as will often have been the case, natural selection will be powerless in certain beneficial directions; but this is no valid objection to its efficiency at other times and in other ways; for we are far from having any reason to suppose that many species ever undergo modification and improvement at the same time
36 in the same area.

## Sexual Selection

Inasmuch as peculiarities often appear under domestication in one sex and become hereditarily attached to that sex, so no doubt it will be under nature. Thus it is rendered possible for the two sexes to be modified through natural selection in relation to different habits of life, as is sometimes the case; or for one sex to be modified in relation to the other sex, as commonly occurs. This leads me to say a few words on what I have called Sexual Selection. This form of selection depends, not on a struggle for existence in relation to other organic beings or to external conditions, but on a struggle between the individuals of one sex, generally the males, for the possession of the other sex. The result is not death to the successful competitor, but few or no offspring. Sexual selection is, therefore, less rigorous than natural selection. Generally, the most vigorous males, those which are best fitted for their places in nature, will leave most progeny. But in many cases, victory depends not so much on general vigour, as on having special weapons, confined to the male sex. A hornless stag or spurless cock would have a poor chance of leaving numerous offspring. Sexual selection, by always allowing the victor to breed, might surely give indomitable courage, length to the spur, and strength to the wing to strike in the spurred leg, in nearly the same manner as does the brutal cockfighter by the careful selection of his best cocks. How low in the scale 37 of nature the law of battle descends, I know not; male alligators have been described as fighting, bellowing, and whirling round, like Indians in a war-dance, for the possession of the females; male salmons have been observed fighting all day long; male stag-beetles sometimes bear wounds from the huge mandibles of other males; the males of certain hymenopterous insects have been frequently seen by that inimitable observer M. Fabre, fighting for a particular female who sits by, an apparently unconcerned beholder of the struggle, and then retires with the conqueror. The war is, perhaps, severest between the males of polygamous animals, and these seem oftenest provided with special weapons. The males of carnivorous animals are already well armed; though to them and to others, special means of defence may be given through means of sexual selection, as the mane of the lion, and the hooked jaw to the male salmon; for the shield may be as important for victory, as the sword or spear.

Amongst birds, the contest is often of a more peaceful character. All those who have attended to the subject, believe that there is the severest rivalry between the males of many species to attract, by singing, the females. The rock-thrush of Guiana, birds of paradise, and some others, congregate; and successive males display with the most elaborate care, and show off in the best manner, their gorgeous plumage; they likewise perform strange
antics before the females, which, standing by as spec-

tators, at last choose the most attractive partner. Those who have closely attended to birds in confinement well know that they often take individual preferences and dislikes: thus Sir R. Heron has described how a pied peacock was eminently attractive to all his hen birds. I cannot here enter on the necessary details; but if man can in a short time give beauty and an elegant carriage to his bantams, according to his standard of beauty, I can see no good reason to doubt that female birds, by selecting, during thousands of generations, the most melodious or beautiful males, according to their standard of beauty, might produce a marked effect. Some well-known laws, with respect to the plumage of male and female birds, in comparison with the plumage of the young, can partly be explained through the action of sexual selection on variations occurring at different ages, and transmitted to the males alone or to both sexes at corresponding ages; but I have not space here to enter on this subject.

Thus it is, as I believe, that when the males and females of any animal have the same general habits of life, but differ in structure, colour, or ornament, such differences have been mainly caused by sexual selection: that is, by individual males having had, in successive generations, some slight advantage over other males, in their weapons, means of defence, or charms, which they have transmitted to their male offspring alone. Yet, I would not wish to attribute all sexual differences to this agency: for we see in our domestic animals peculiarities arising and becoming attached to the male sex, which apparently have not been augmented through selection by man. The tuft of hair on the breast of the wild turkey-cock cannot be of any use, and it is doubtful whether it can be ornamental in the eyes of the female bird; – indeed, had the tuft appeared under domestication, it would have been called a monstrosity.

## Illustrations of the Action of Natural Selection, or the Survival of the Fittest

In order to make it clear how, as I believe, natural selection acts, I must beg permission to give one or two imaginary illustrations. Let us take the case of a wolf, which preys on various animals, securing some by craft, some by strength, and some by fleetness; and let us suppose that the fleetest prey, a deer for instance, had from any change in the country increased in numbers, or that other prey had decreased in numbers, during that season of the year when the wolf was hardest pressed for food. Under such circumstances the swiftest and slimmest wolves would have the best chance of surviving and so be preserved or selected, – provided always that they retained strength to master their prey at this or some other period of the year, when they were compelled to prey on other animals. I can see no more reason to doubt that this would be the result, than that man should be able to improve the fleetness of his greyhounds by careful and methodical selection, or by that kind of unconscious selection which follows from each man trying to keep the best dogs without any thought of modifying the breed. I may add, that, according to Mr Pierce, there are two varieties of the wolf inhabiting the Catskill Mountains, in the United States, one with a light greyhound-like form, which pursues deer, and the other more bulky, with shorter legs, which more frequently attacks the shepherd's flocks.

It should be observed that, in the above illustration, I speak of the slimmest individual wolves, and not of any single strongly-marked variation having been preserved. In former editions of this work I sometimes spoke as if this latter alternative had frequently occurred. I saw the great importance of individual differences, and this led me fully to discuss the results of unconscious selection by man, which depends on the preservation of all the more or less valuable individuals, and on the destruction of the worst. I saw, also, that the preservation in a state of nature of any occasional deviation of structure, such as a monstrosity, would be a rare event; and that, if at first preserved, it would generally be lost by subsequent intercrossing with ordinary individuals. Nevertheless, until reading an able and valuable article in the 'North British Review' (1867), I did not appreciate how rarely single variations, whether slight or strongly-marked, could be perpetuated. The author takes the case of a pair of animals, producing during their lifetime two hundred 41 offspring, of which, from various causes of destruction, only two on an average survive to pro-create their kind. This is rather an extreme estimate for most of the higher animals, but by no means so for many of the lower organisms. He then shows that if a single individual were born, which varied in some manner, giving it twice as good a chance of life as that of the other individuals, yet the chances would be strongly against its survival. Supposing it to survive and to breed, and that half its young inherited the favourable variation; still, as the Reviewer goes on to show, the young would have only a slightly better chance of surviving and breeding; and this chance would go on decreasing in the succeeding generations. The justice of these remarks cannot, I think, be disputed. If, for instance, a bird of some kind could procure its food more easily by having its beak curved, and if one were born with its beak strongly curved, and which consequently flourished, nevertheless there would be a very poor chance of this one individual perpetuating its kind to the exclusion of the common form; but there can hardly be a doubt, judging by what we see taking place under domestication, that this result would follow from the preservation during many generations of a large number of individuals with more or less strongly curved beaks, and from the destruction of a still larger number with the straightest beaks.

It should not, however, be overlooked that certain
rather strongly marked variations, which no one would

rank as mere individual differences, frequently recur owing to a similar organisation being similarly acted on – of which fact numerous instances could be given with our domestic productions. In such cases, if the varying individual did not actually transmit to its offspring its newly-acquired character, it would undoubtedly transmit to them, as long as the existing conditions remained the same, a still stronger tendency to vary in the same manner. There can also be little doubt that the tendency to vary in the same manner has often been so strong that all the individuals of the same species have been similarly modified without the aid of any form of selection. Or only a third, fifth, or tenth part of the individuals may have been thus affected, of which fact several instances could be given. Thus Graba estimates that about one-fifth of the guillemots in the Faroe Islands consist of a variety so well marked, that it was formerly ranked as a distinct species under the name of Uria lacrymans. In cases of this kind, if the variation were of a beneficial nature, the original form would soon be supplanted by the modified form, through the survival of the fittest.

To the effects of intercrossing in eliminating variations of all kinds, I shall have to recur; but it may be here remarked that most animals and plants keep to their proper homes, and do not needlessly wander about; we see this even with migratory birds, which almost always return to the same spot. Consequently each newly-formed variety would generally be at first local, as seems to be the common rule with varieties in a state of nature; so that similarly modified individuals would soon exist in a small body together, and would often breed together. If the new variety were successful in its battle for life, it would slowly spread from a central district, competing with and conquering the unchanged individuals on the margins of an ever-increasing circle.

It may be worth while to give another and more complex illustration of the action of natural selection. Certain plants excrete sweet juice, apparently for the sake of eliminating something injurious from the sap: this is effected, for instance, by glands at the base of the stipules in some Leguminosæ, and at the backs of the leaves of the common laurel. This juice, though small in quantity, is greedily sought by insects; but their visits do not in any way benefit the plant. Now, let us suppose that the juice or nectar was excreted from the inside of the flowers of a certain number of plants of any species. Insects in seeking the nectar would get dusted with pollen, and would often transport it from one flower to another. The flowers of two distinct individuals of the same species would thus get crossed; and the act of crossing, as can be fully proved, gives rise to vigorous seedlings, which consequently would have the best chance of flourishing and surviving. The plants which produced flowers with the largest glands or nectaries, excreting most nectar, would oftenest be visited by insects, and would oftenest be crossed; and so in the long-run would gain the upper hand and form a local variety. The flowers, also, which had their stamens and pistils placed, in relation to the size and habits of the particular insect which visited them, so as to favour in any degree the transportal of the pollen, would likewise be favoured. We might have taken the case of insects visiting flowers for the sake of collecting pollen instead of nectar; and as pollen is formed for the sole purpose of fertilisation, its destruction appears to be a simple loss to the plant; yet if a little pollen were carried, at first occasionally and then habitually, by the pollen-devouring insects from flower to flower, and a cross thus effected, although nine-tenths of the pollen were destroyed it might still be a great gain to the plant to be thus robbed; and the individuals which produced more and more pollen, and had larger anthers, would be selected.

When our plant, by the above process long continued, had been rendered highly attractive to insects, they would, unintentionally on their part, regularly carry pollen from flower to flower; and that they do this effectually, I could easily show by many striking facts. I will give only one, as likewise illustrating one step in the separation of the sexes of plants. Some holly-trees bear only male flowers, which have four stamens producing a rather small quantity of pollen, and a rudimentary pistil; other holly-trees bear only female flowers; these have a full-sized pistil, and four stamens with shrivelled anthers, in which not a grain of pollen can be detected. Having found a female tree exactly sixty yards from a male tree,

I put the stigmas of twenty flowers, taken from different branches, under the microscope, and on all, without exception, there were a few pollen-grains, and on some a profusion. As the wind had set for several days from the female to the male tree, the pollen could not thus have been carried. The weather had been cold and boisterous, and therefore not favourable to bees, nevertheless every female flower which I examined had been effectually fertilised by the bees, which had flown from tree to tree in search of nectar. But to return to our imaginary case: as soon as the plant had been rendered so highly attractive to insects that pollen was regularly carried from flower to flower, another process might commence. No naturalist doubts the advantage of what has been called the 'physiological division of labour;' hence we may believe that it would be advantageous to a plant to produce stamens alone in one flower or on one whole plant, and pistils alone in another flower or on another plant. In plants under culture and placed under new conditions of life, sometimes the male organs and sometimes the female organs become more or less impotent; now if we suppose this to occur in ever so slight a degree under nature, then, as pollen is already carried regularly from flower to flower, and as a more complete separation of the sexes of our plant would be advantageous on the principle of the division of labour, individuals with this tendency more and more increased, would be continually

favoured or selected, until at last a complete separation

of the sexes might be effected. It would take up too much space to show the various steps, through dimorphism and other means, by which the separation of the sexes in plants of various kinds is apparently now in progress; but I may add that some of the species of holly in North America, are, according to Asa Gray, in an exactly intermediate condition, or, as he expresses it, are more or less diœciously polygamous.

Let us now turn to the nectar-feeding insects; we may suppose the plant, of which we have been slowly increasing the nectar by continued selection, to be a common plant; and that certain insects depended in main part on its nectar for food. I could give many facts showing how anxious bees are to save time: for instance, their habit of cutting holes and sucking the nectar at the bases of certain flowers, which with a very little more trouble, they can enter by the mouth. Bearing such facts in mind, it may be believed that under certain circumstances individual differences in the curvature or length of the proboscis, &c., too slight to be appreciated by us, might profit a bee or other insect, so that certain individuals would be able to obtain their food more quickly than others; and thus the communities to which they belonged would flourish and throw off many swarms inheriting the same peculiarities. The tubes of the corolla of the common red and incarnate clovers (Trifolium pratense and incarnatum) do not on a hasty glance appear to differ in length; yet the hive-bee can easily suck the nectar out 47 of the incarnate clover, but not out of the common red clover, which is visited by humble-bees alone; so that whole fields of the red clover offer in vain an abundant supply of precious nectar to the hive-bee. That this nectar is much liked by the hive-bee is certain; for I have repeatedly seen, but only in the autumn, many hive-bees sucking the flowers through holes bitten in the base of the tube by humble-bees. The difference in the length of the corolla in the two kinds of clover, which determines the visits of the hive-bee, must be very trifling; for I have been assured that when red clover has been mown, the flowers of the second crop are somewhat smaller, and that these are visited by many hive-bees. I do not know whether this statement is accurate; nor whether another published statement can be trusted, namely, that the Ligurian bee, which is generally considered a mere variety of the common hive-bee, and which freely crosses with it, is able to reach and suck the nectar of the red clover. Thus, in a country where this kind of clover abounded, it might be a great advantage to the hive-bee to have a slightly longer or differently constructed proboscis. On the other hand, as the fertility of this clover absolutely depends on bees visiting the flowers, if humble-bees were to become rare in any country, it might be a great advantage to the plant to have a shorter or more deeply divided corolla, so that the hive-bees should be enabled to suck its flowers. Thus I can understand how a flower

and a bee might slowly become, either simultaneously or

one after the other, modified and adapted to each other in the most perfect manner, by the continued preservation of all the individuals which presented slight deviations of structure mutually favourable to each other.

I am well aware that this doctrine of natural selection, exemplified in the above imaginary instances, is open to the same objections which were first urged against Sir Charles Lyell's noble views on 'the modern changes of the earth, as illustrative of geology;' but we now seldom hear the agencies which we see still at work, spoken of as trifling or insignificant, when used in explaining the excavation of the deepest valleys or the formation of long lines of inland cliffs. Natural selection acts only by the preservation and accumulation of small inherited modifications, each profitable to the preserved being; and as modern geology has almost banished such views as the excavation of a great valley by a single diluvial wave, so will natural selection banish the belief of the continued creation of new organic beings, or of any great and sudden modification in their structure . . .

*Circumstances favourable for the production of new forms through Natural Selection*

This is an extremely intricate subject. A great amount of variability, under which term individual differences are always included, will evidently be favourable. A large number of individuals, by giving a better chance within any given period for the appearance of profitable variations, will compensate for a lesser amount of variability in each individual, and is, I believe, a highly important element of success. Though Nature grants long periods of time for the work of natural selection, she does not grant an indefinite period; for as all organic beings are striving to seize on each place in the economy of nature, if any one species does not become modified and improved in a corresponding degree with its competitors, it will be exterminated. Unless favourable variations be inherited by some at least of the offspring, nothing can be effected by natural selection. The tendency to reversion may often check or prevent the work; but as this tendency has not prevented man from forming by selection numerous domestic races, why should it prevail against natural selection?

In the case of methodical selection, a breeder selects for some definite object, and if the individuals be allowed freely to intercross, his work will completely fail. But when many men, without intending to alter the breed, have a nearly common standard of perfection, and all try to procure and breed from the best animals, improvement surely but slowly follows from this unconscious process of selection, notwithstanding that there is no separation of selected individuals. Thus it will be under nature; for within a confined area, with some place in the natural polity not perfectly occupied, all the individuals varying in the right direction, though in different degrees, will tend to be preserved. But if the area be large, its several districts will almost certainly present different conditions of life; and then, if the same species undergoes modification in different districts, the newly-formed varieties will intercross on the confines of each. But we shall see in the sixth chapter that intermediate varieties, inhabiting intermediate districts, will in the long run generally be supplanted by one of the adjoining varieties. Intercrossing will chiefly affect those animals which unite for each birth and wander much, and which do not breed at a very quick rate. Hence with animals of this nature, for instance, birds, varieties will generally be confined to separated countries; and this I find to be the case. With hermaphrodite organisms which cross only occasionally, and likewise with animals which unite for each birth, but which wander little and can increase at a rapid rate, a new and improved variety might be quickly formed on any one spot, and might there maintain itself in a body and afterwards spread, so that the individuals of the new variety would chiefly cross together. On this principle, nurserymen always prefer saving seed from a large body of plants, as the chance of intercrossing is thus lessened.

Even with animals which unite for each birth, and which do not propagate rapidly, we must not assume that free intercrossing would always eliminate the effects of natural selection; for I can bring forward a considerable body of facts showing that within the same area, two varieties of the same animal may long remain 51 distinct, from haunting different stations, from breeding at slightly different seasons, or from the individuals of each variety preferring to pair together.

Intercrossing plays a very important part in nature by keeping the individuals of the same species, or of the same variety, true and uniform in character. It will obviously thus act far more efficiently with those animals which unite for each birth; but, as already stated, we have reason to believe that occasional intercrosses take place with all animals and plants. Even if these take place only at long intervals of time, the young thus produced will gain so much in vigour and fertility over the offspring from long-continued self-fertilisation, that they will have a better chance of surviving and propagating their kind; and thus in the long run the influence of crosses, even at rare intervals, will be great. With respect to organic beings extremely low in the scale, which do not propagate sexually, nor conjugate, and which cannot possibly intercross, uniformity of character can be retained by them under the same conditions of life, only through the principle of inheritance, and through natural selection which will destroy any individuals departing from the proper type. If the conditions of life change and the form undergoes modification, uniformity of character can be given to the modified offspring, solely by natural selection preserving similar favourable variations.

Isolation, also, is an important element in the modification of species through natural selection. In a confined or isolated area, if not very large, the organic and inorganic conditions of life will generally be almost uniform; so that natural selection will tend to modify all the varying individuals of the same species in the same manner. Intercrossing with the inhabitants of the surrounding districts will, also, be thus prevented. Moritz Wagner has lately published an interesting essay on this subject, and has shown that the service rendered by isolation in preventing crosses between newly-formed varieties is probably greater even than I supposed. But from reasons already assigned I can by no means agree with this naturalist, that migration and isolation are necessary elements for the formation of new species. The importance of isolation is likewise great in preventing, after any physical change in the conditions, such as of climate, elevation of the land, &c., the immigration of better adapted organisms; and thus new places in the natural economy of the district will be left open to be filled up by the modification of the old inhabitants. Lastly, isolation will give time for a new variety to be improved at a slow rate; and this may sometimes be of much importance. If, however, an isolated area be very small, either from being surrounded by barriers, or from having very peculiar physical conditions, the total number of the inhabitants will be small; and this will retard the production of new species through natural selection, by decreasing the chances of favourable variations arising.

The mere lapse of time by itself does nothing, either for or against natural selection. I state this because it has been erroneously asserted that the element of time has been assumed by me to play an all-important part in modifying species, as if all the forms of life were necessarily undergoing change through some innate law. Lapse of time is only so far important, and its importance in this respect is great, that it gives a better chance of beneficial variations arising and of their being selected, accumulated, and fixed. It likewise tends to increase the direct action of the physical conditions of life, in relation to the constitution of each organism.

If we turn to nature to test the truth of these remarks, and look at any small isolated area, such as an oceanic island, although the number of species inhabiting it is small, as we shall see in our chapter on Geographical Distribution; yet of these species a very large proportion are endemic, – that is, have been produced there and nowhere else in the world. Hence an oceanic island at first sight seems to have been highly favourable for the production of new species. But we may thus deceive ourselves, for to ascertain whether a small isolated area, or a large open area like a continent, has been most favourable for the production of new organic forms, we ought to make the comparison within equal times; and this we are incapable of doing.

Although isolation is of great importance in the production of new species, on the whole I am inclined to believe that largeness of area is still more important, especially for the production of species which shall prove capable of enduring for a long period, and of spreading widely. Throughout a great and open area, not only will there be a better chance of favourable variations, arising from the large number of individuals of the same species there supported, but the conditions of life are much more complex from the large number of already existing species; and if some of these many species become modified and improved, others will have to be improved in a corresponding degree, or they will be exterminated. Each new form, also, as soon as it has been much improved, will be able to spread over the open and continuous area, and will thus come into competition with many other forms. Moreover, great areas, though now continuous, will often, owing to former oscillations of level, have existed in a broken condition; so that the good effects of isolation will generally, to a certain extent, have concurred. Finally, I conclude that, although small isolated areas have been in some respects highly favourable for the production of new species, yet that the course of modification will generally have been more rapid on large areas; and what is more important, that the new forms produced on large areas, which already have been victorious over many competitors, will be those that will spread most widely, and will give rise to the greatest number of new varieties and species. They will thus play a more important part in the changing history of the organic world.

In accordance with this view, we can, perhaps, understand some facts which will be again alluded to in our chapter on Geographical Distribution; for instance, the fact of the productions of the smaller continent of Australia now yielding before those of the larger Europæo-Asiatic area. Thus, also, it is that continental productions have everywhere become so largely naturalised on islands. On a small island, the race for life will have been less severe, and there will have been less modification and less extermination. Hence, we can understand how it is that the flora of Madeira, according to Oswald Heer, resembles to a certain extent the extinct tertiary flora of Europe. All fresh-water basins, taken together, make a small area compared with that of the sea or of the land. Consequently, the competition between fresh-water productions will have been less severe than elsewhere; new forms will have been then more slowly produced, and old forms more slowly exterminated. And it is in fresh-water basins that we find seven genera of Ganoid fishes, remnants of a once preponderant order: and in fresh water we find some of the most anomalous forms now known in the world as the Ornithorhynchus and Lepidosiren, which, like fossils, connect to a certain extent orders at present widely sundered in the natural scale. These anomalous forms may be called living fossils; they have endured to the present day, from having inhabited a confined area, and from having been exposed to less varied, and therefore less severe, competition.

To sum up, as far as the extreme intricacy of the subject permits, the circumstances favourable and unfavourable for the production of new species through natural selection. I conclude that for terrestrial productions a large continental area, which has undergone many oscillations of level, will have been the most favourable for the production of many new forms of life, fitted to endure for a long time and to spread widely. Whilst the area existed as a continent, the inhabitants will have been numerous in individuals and kinds, and will have been subjected to severe competition. When converted by subsidence into large separate islands, there will still have existed many individuals of the same species on each island: intercrossing on the confines of the range of each new species will have been checked: after physical changes of any kind, immigration will have been prevented, so that new places in the polity of each island will have had to be filled up by the modification of the old inhabitants; and time will have been allowed for the varieties in each to become well modified and perfected. When, by renewed elevation, the islands were reconverted into a continental area, there will again have been very severe competition: the most favoured or improved varieties will have been enabled to spread: there will have been much extinction of the less improved forms, and the relative porportional numbers of the various inhabitants of the reunited continent will again have been changed; and again there will have been a fair field for natural selection to improve still further the inhabitants, and thus to produce new species.

That natural selection generally acts with extreme slowness I fully admit. It can act only when there are places in the natural polity of a district which can be better occupied by the modification of some of its existing inhabitants. The occurrence of such places will often depend on physical changes, which generally take place very slowly, and on the immigration of better adapted forms being prevented. As some few of the old inhabitants become modified, the mutual relations of others will often be disturbed; and this will create new places, ready to be filled up by better adapted forms; but all this will take place very slowly. Although all the individuals of the same species differ in some slight degree from each other, it would often be long before differences of the right nature in various parts of the organisation might occur. The result would often be greatly retarded by free intercrossing. Many will exclaim that these several causes are amply sufficient to neutralise the power of natural selection. I do not believe so. But I do believe that natural selection will generally act very slowly, only at long intervals of time, and only on a few of the inhabitants of the same region. I further believe that these slow, intermittent results accord well with what geology tells us of the rate and manner at which the inhabitants of the world have changed.

Slow though the process of selection may be, if feeble

man can do much by artificial selection, I can see no limit to the amount of change, to the beauty and complexity of the coadaptations between all organic beings, one with another and with their physical conditions of life, which may have been affected in the long course of time through nature's power of selection, that is by the survival of the fittest . . .